Members of the Animal Kingdom

AMPHIBIANS

TRACY VONDER BRINK

A Crabtree Crown Book

Crabtree Publishing
crabtreebooks.com

School-to-Home Support for Caregivers and Teachers

This appealing book is designed to teach students about core subject areas. Students will build upon what they already know about the subject, and engage in topics that they want to learn more about. Here are a few guiding questions to help readers build their comprehensions skills. Possible answers appear here in red.

Before Reading:

What do I know about mammals?

- *I know amphibians are animals.*
- *I know a frog is an amphibian.*

What do I want to learn about this topic?

- *I want to learn about where amphibians live.*
- *I want to learn how amphibians are different from other animals.*

During Reading:

I'm curious to know...

- *I'm curious to know how a tadpole becomes a frog.*
- *I'm curious to know how amphibians hunt.*

How is this like something I already know?

- *I know that tadpoles grow into frogs.*
- *I know that some amphibians eat bugs.*

After Reading:

What was the author trying to teach me?

- *The author was trying to teach me how amphibians are different from other animals.*
- *The author was trying to teach me what metamorphosis is.*

How did the photographs and captions help me understand more?

- *The photographs helped me understand what different amphibians look like.*
- *The captions gave me extra information about amphibian body parts.*

TABLE OF CONTENTS

Chapter 1

SORTING IT OUT

People have been curious about living things for thousands of years. One way to study living things is to organize them into groups. Scientists call this **classification**. Living creatures are classified based on the **traits** they have in common.

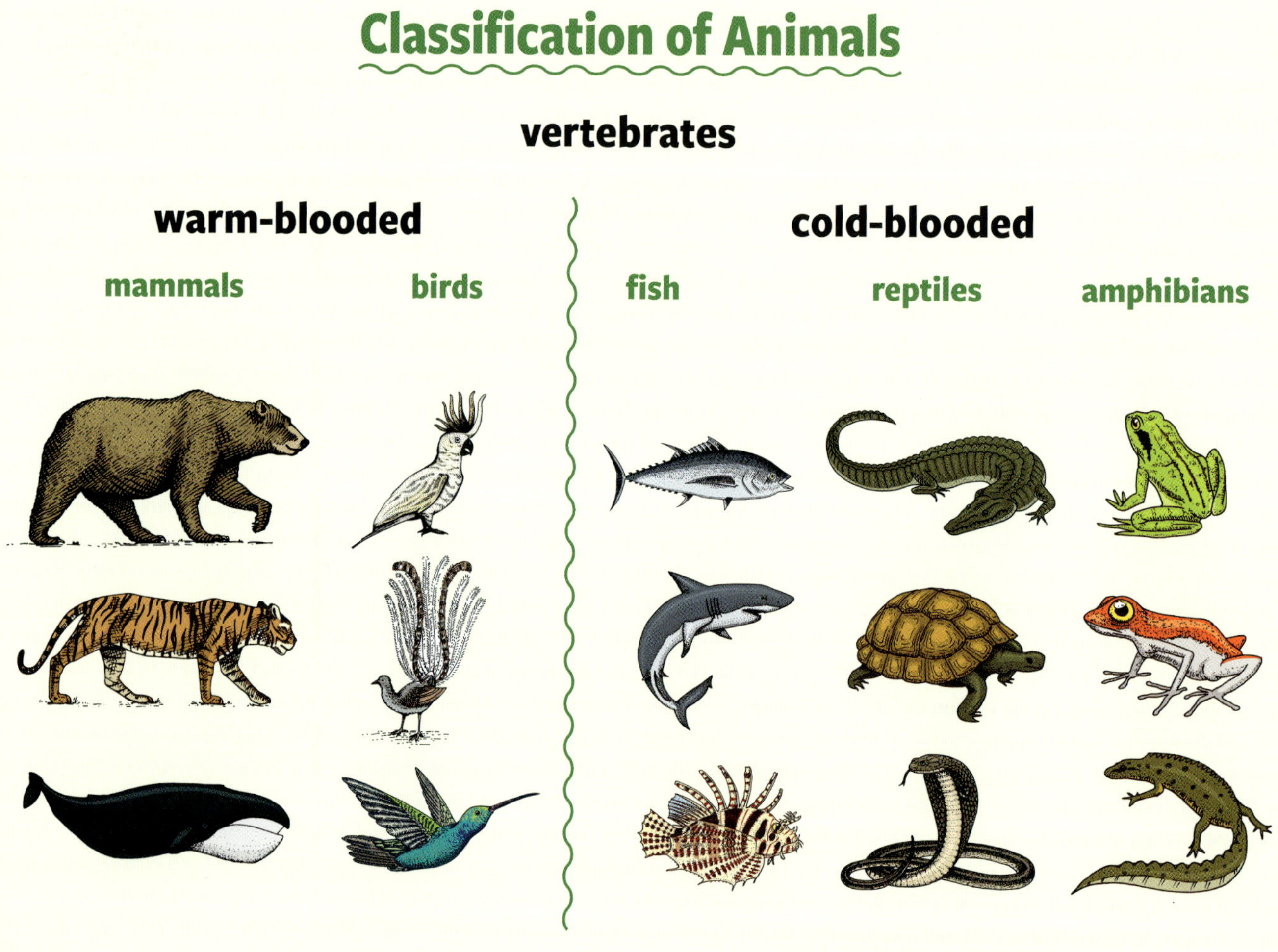

Animals are classified in two main groups. Vertebrates have backbones. Invertebrates do not have backbones.

Living things are divided into kingdoms, such as the animal kingdom. A kingdom contains groups of many living creatures. Amphibians are one group in the animal kingdom. Each group in a kingdom is then divided into smaller groups. For example, frogs, salamanders, and caecilians are groups of amphibians.

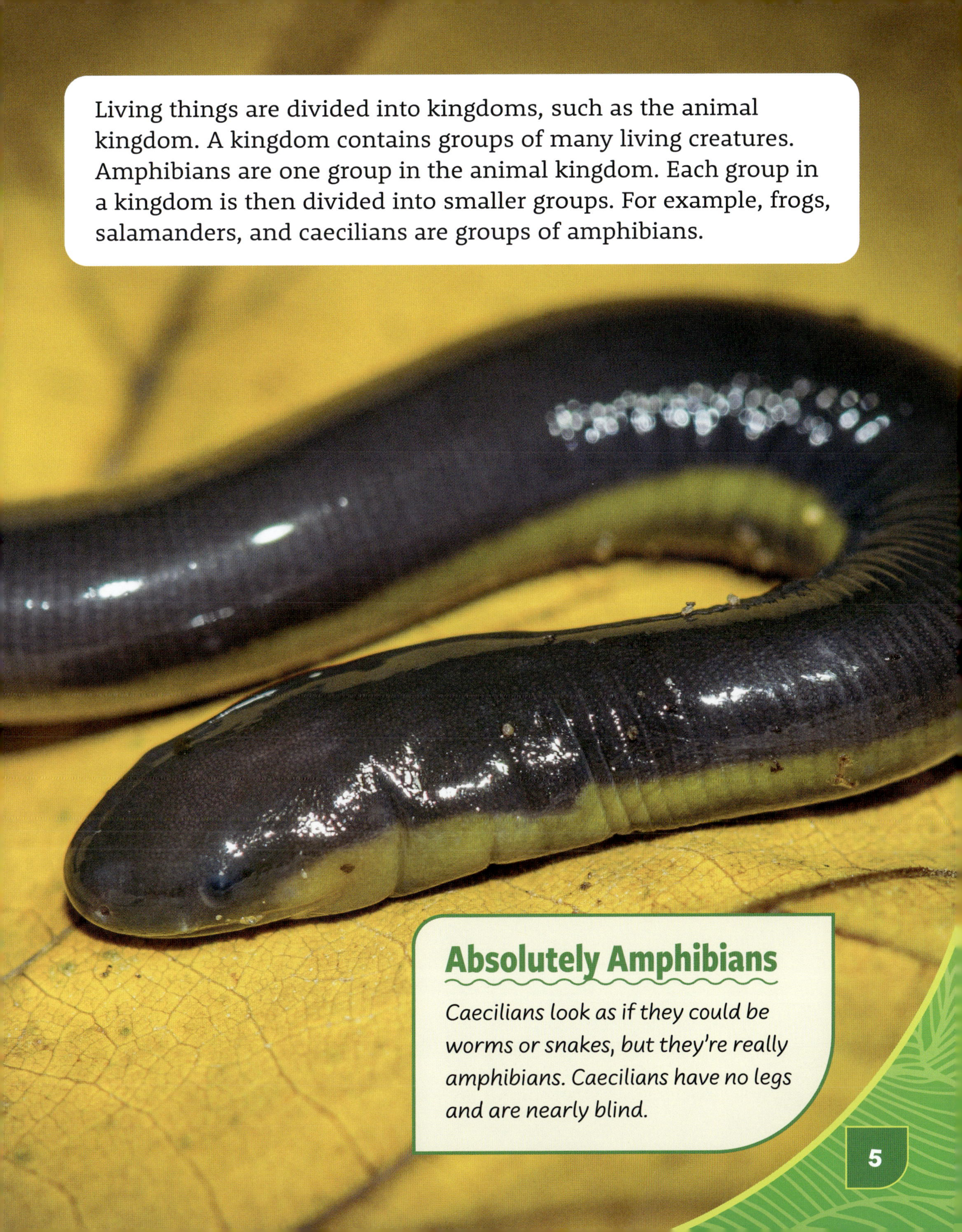

Absolutely Amphibians

Caecilians look as if they could be worms or snakes, but they're really amphibians. Caecilians have no legs and are nearly blind.

A species is the most specific way to classify animals. There are many species within a group. Members of the same species share the same traits and are able to have babies with each other. The green frog is one amphibian species.

green frog

CHAPTER 2

MEET THE AMPHIBIANS

The name amphibian comes from Greek words meaning “a double life.” This animal group was given that name because most amphibian species begin their lives in water. Then, they go through **metamorphosis** and live on land as adults. Some amphibian adults live both on land and in water.

Amphibians are cold-blooded. That means they can't control their own body temperature. Instead, they depend on their environment to **regulate** it. They stay in the Sun until they are warm enough to move, hunt, and eat. Amphibians cool down by moving into the shade or water.

green frog

Absolutely Amphibians

Some amphibians that live in areas with cold weather ***hibernate*** *when the temperature drops. American toads (left) dig down about 2 feet (.61 m) and spend winters tucked into their burrows.*

Amphibians have thin skin. Water and **oxygen** pass through their skin and into their bodies. So, amphibians breathe air through their skin as well as with their lungs. Many species of salamanders have no lungs and breathe only through their skin.

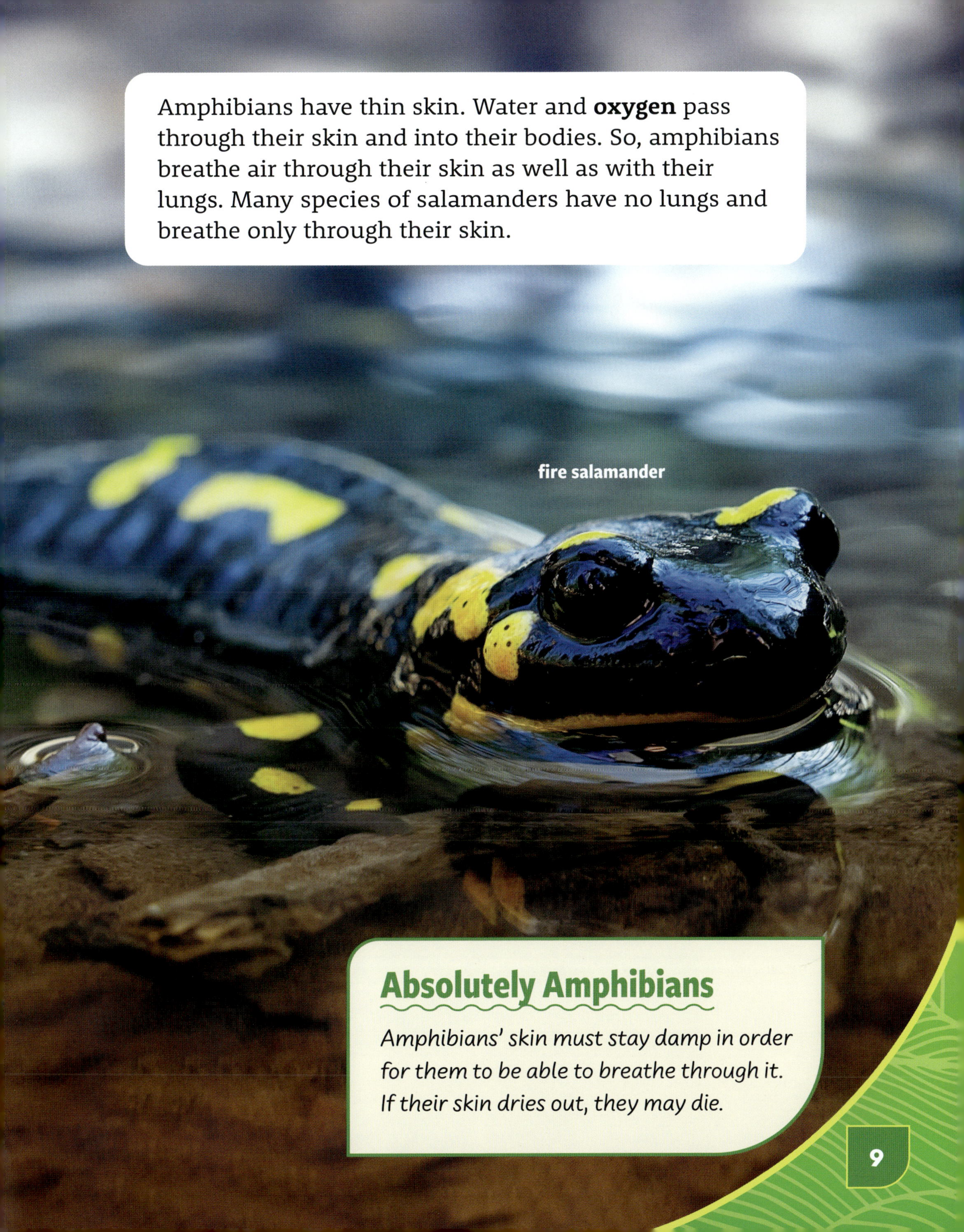

fire salamander

Absolutely Amphibians

Amphibians' skin must stay damp in order for them to be able to breathe through it. If their skin dries out, they may die.

Amphibians are vertebrates. That means they have backbones. But amphibian backbones are not all the same. Frogs have short, stiff **spines** built to handle jumping. Salamanders have backbones that bend as they walk. Caecilians have as many as 285 bones in their spines.

Carolina Sandhills salamander

CHAPTER 3

HOME SWEET HOME

Amphibians are found on every continent except Antarctica. Because amphibians need to keep their skin damp, many live near water in tropical or **temperate** regions. A wide variety of frog and toad species are found in tropical rain forests.

red-eyed tree frog

Absolutely Amphibians

People have built on and changed many of the areas where amphibians need to live. Today, about 40% of amphibian species are at risk of extinction.

FROGS AND TOADS

Frogs live near freshwater lakes, ponds, rivers, and swamps all over the world. Toads are frogs, but toads have drier skin and don't need as much water. Toads are found in forests and grasslands, as well as around bodies of water. They can even be found in deserts.

Some kinds of frogs have webbed feet to help them swim. Most toads don't swim as much as frogs, but the Colorado river toad (right) spends some of its time in water.

SALAMANDERS

Many salamander species also live in or near water. On land, they often live in shady forests where it is easier to stay cool and moist. Some salamander species burrow into soil. Spotted salamanders hide under rocks and logs in the woodlands of eastern North America.

spotted salamander

Absolutely Amphibians

Salamanders may look like lizards, but lizards are reptiles. Salamanders are amphibians. Salamanders have smooth skin, and lizards have scales.

CAECILIANS

Caecilians live in tropical areas around the world. Caecilians spend their lives underground. Some species live in shallow streams. Caecilians are rarely seen by people.

female caecilian with her young

CHAPTER 4

GROWING UP

Amphibian eggs are soft. Most amphibians lay their eggs in water or somewhere damp so the eggs don't dry out. Many lay hundreds of eggs at a time. An amphibian **embryo** grows inside the egg.

agile frog on eggs

Absolutely Amphibians

Many amphibian parents leave their eggs alone. They do not stay to care for the babies. But one kind of caecilian young stay with their mother for four to six weeks. They feed by eating bits of her skin.

LARVA

The embryo grows into a **larva** and hatches from the egg. The larva of a frog or toad is called a tadpole. Tadpoles live underwater, breathe through gills, and have a tail. Many salamander larvae also do this.

tree frog tadpole

eggs of the Golden-striped salamander

Absolutely Amphibians

Sirens are a kind of amphibian that keep their gills and spend their lives underwater. They become adults without going through metamorphosis.

METAMORPHOSIS

Most amphibian larvae go through metamorphosis to become adults. This takes most species about two to four months. During metamorphosis, amphibians develop lungs to breathe air. Tadpoles lose their tails and grow legs. Salamanders keep their tails and grow legs. After metamorphosis, amphibians are able to live on land and find mates.

green frog tadpole

Absolutely Amphibians

Compared to frogs and salamanders, caecilians do not change as much during metamorphosis. They do lose their gills and grow lungs, however.

CHAPTER 5

ON THE MOVE

Toads have short legs. They move by crawling or with short hops. Toads also have claw-like fingers they can use to grip roots and stumps as they crawl. They usually stay on the ground and don't often climb.

Touching a toad won't give you warts, but toads can make a protective liquid that irritates human skin.

Frogs have long back legs built for leaping. Before jumping, their leg muscles shorten and load energy into their **tendons**. Then the tendons release like a spring. This lets frogs jump far. The South African sharp-nosed frog is only 3 inches (7.6 cm) long but can jump more than 9 feet (3 meters)!

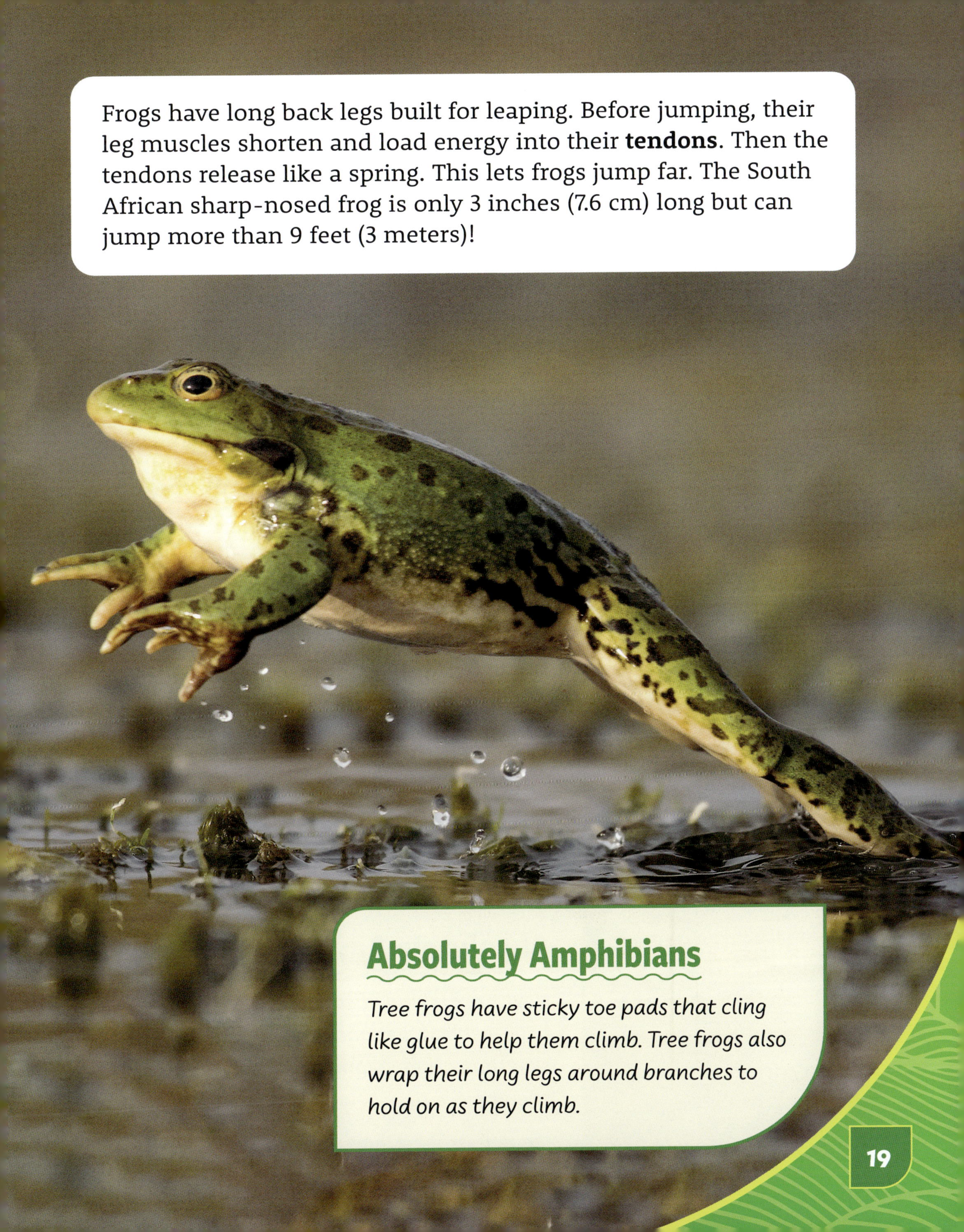

Absolutely Amphibians

Tree frogs have sticky toe pads that cling like glue to help them climb. Tree frogs also wrap their long legs around branches to hold on as they climb.

Many species of salamanders have four legs, but some have two. Salamanders are low to the ground. They move by bending their bodies from side to side as they crawl. They usually move slowly, but will quickly scuttle away from danger.

Absolutely Amphibians

Newts are a kind of salamander. Newts have webbed feet and tend to have rougher skin than other salamanders. They often spend more time in water than on land.

Caecilians are burrowers. They use their strong, thick skulls to push through soil. They scrunch up their backbones and pull their bodies forward. Some caecilians swim underwater and come to the surface to breathe.

Congo caecilian

CHAPTER 6

ON THE HUNT

FROGS AND TOADS

Frogs and toads have sticky tongues. They use their tongues to snatch bugs, snails, and worms from the ground and flies and moths out of the air. They pull the prey into their mouths and swallow them whole.

tree frog

SALAMANDERS

Salamanders have small teeth in their upper and lower jaws. They snap up worms, bugs, and even other salamanders. Salamanders that live in water hunt small fish, frogs or tadpoles.

smallmouth salamander

CAECILIANS

Caecilians have dozens of sharp, needle-like teeth. They eat worms, bugs, small snakes, and other caecilians. When prey comes near their burrows, caecilians snatch it with their teeth. Then they spin to pull the prey into their burrow.

Caecilian teeth are made for grabbing but not chewing. They swallow their prey whole.

CHAPTER 7

AMAZING AMPHIBIANS

AXOLOTL

Axolotls never lose their gills. They spend their lives underwater in a lake and waterways in the southern part of Mexico City. They are able to regrow lost legs and can even regrow damaged hearts, spines, and brains.

Because they keep their gills, axolotls don't go through metamorphosis.

VIETNAMESE MOSSY FROG

Vietnamese mossy frogs live in caves and near mountain streams in northern Vietnam. Their color and bumpy skin makes them look like clumps of moss. Mossy frogs sometimes roll into a ball and play dead when frightened.

Vietnamese mossy frogs hide in water. They also attach themselves to rocks to look like clumps of moss.

CHINESE GIANT SALAMANDER

Chinese giant salamanders can grow nearly 6 feet (1.8 meters) long. They live in rivers and streams in China. They spend their lives underwater but don't have gills, so they come to the surface to breathe. These large amphibians live around 80 years in the wild.

Chinese giant salamanders live the longest of any amphibian species.

GLASS FROG

Glass frogs usually live in the rain forests of southern Mexico, Central America, and South America. The skin on their bellies and chests is see-through. It's possible to watch their hearts beating or food moving through their bodies.

Glass frogs are see-through as part of their camouflage. It helps the frogs blend in with the color of its surroundings.

Scientists have found more than 8,000 species of amphibians worldwide. New species are found every year. How many amphibians can you name?

The southern banded newt uses the brightly colored crests on his back and tail to attract females.

GLOSSARY

classification (klas-uh-fuh-KAY-shun): The arrangement of things into groups according to a system

embryo (EM-bree-oh): A living creature in the earliest stage of development, before it is hatched or born

hibernate (HI-ber-neyt): To spend the winter in a dormant, or inactive, state

larva (LAR-vuh): A stage in the development in some animals after birth or hatching and before metamorphosis

metamorphosis (meh-tuh-MOR-fuh-sis): A major change in the form of some living creatures that happens as they become adults

oxygen (AHK-suh-jn): A gas, present in air and water, that all living things need to survive

regulate (REG-yuh-late): To adjust the amount of something, such as heat

spine (SPYN): The row of connected bones down the middle of the back

temperate (TEM-pr-uht): Neither very hot nor very cold

tendon (TEN-duhn): The cord of tough fiber that attaches a muscle to a bone

traits (TRAYTS): Qualities or characteristics that make one thing different from another

INDEX

COMPREHENSION QUESTIONS

1. Amphibians are vertebrates. That means they have____.

a. gills

b. backbones

c. eggs

2. Why do many amphibians live near water?

a. They must keep their skin damp.

b. They drink a lot of water daily.

c. They can only sleep in damp soil.

3. The larva of a frog is called a _____.

a. toad

b. embryo

c. tadpole

4. True or False: Amphibians are cold-blooded.

5. True or False: Toads are not frogs.

Answers: 1. B, 2. A, 3. C, 4. True, 5. False

ABOUT THE AUTHOR

Tracy Vonder Brink loves to learn about science and nature. She has written many nonfiction books for kids and is a contributing editor for three children's science magazines. Tracy lives in Cincinnati, Ohio with her husband, two daughters, and two rescue dogs.

Written by: Tracy Vonder Brink
Designed by: Kathy Walsh
Series Development: James Earley
Proofreader: Janine Deschenes
Educational Consultant: Marie Lemke M.Ed.

Photographs: Shutterstock; Cover: Eugenia Petrovskaya, LorraineHudgins, Ken Griffiths, Lauren Suryanata, Milan Zygmunt, Beatrice Prezzemoli; p 2- 31 Eugenia Petrovskaya; p 4 Arthur Balitskii; p 5 WildWoodMan; p 6 LorraineHudgins; p 7 W. de Vries; p 8 Gerry Bishop; p 9 Rejdan; p 10 Nathan A Shepard; p 11 Vaclav Sebek; p 12 manfredxy, Amy Newton-McConnel; p 13 Giulio Benzin; p 14 The Natural History Museum, London/Science Source; p 15 Matteo photos; p 16 W. de Vries, Taniaaraujo; p 17 Ken Griffiths; p 18 John Arehart; p 19 WildlifeWorld; p 20 MargieV; p 21 Dante Fenolio/ Science Source; p 22 Cathy Keifer; p 23 Mike Wilhelm; p 24 Danté Fenolio/Science Source; p 25 Arm001; p 26 davemhuntphotography; p 27 tristan tan; p 28 Dr Morley Read; p 29 Dirk Ercken, Rudmer Zwerver

Crabtree Publishing

crabtreebooks.com 800-387-7650

Printed in the U.S.A./012023/CG20220815

Published in Canada Crabtree Publishing
616 Welland Ave.
St. Catharines, Ontario
L2M 5V6

Published in the United States Crabtree Publishing
347 Fifth Ave
Suite 1402-145
New York, NY 10016

Library and Archives Canada Cataloguing in Publication
Available at Library and Archives Canada

Library of Congress Cataloging-in-Publication Data
Available at the Library of Congress

Hardcover: 978-1-0398-0022-9
Paperback: 978-1-0398-0081-6
Ebook (pdf): 978-1-0398-0200-1
Epub: 978-1-0398-0140-0